Appi Alex

Gives App Advice

AF808664

Story by:
Lemi-Ola Erinkitola

Illustrated by:
Yuffie Yuliana

www.thecriticalthinkingchild.com/cybersafety
ISBN 978-0-9899933-8-8

The fourth character introduced in the Cyber City book series is **Appi Alex**, who takes the form of a phone app.

He is best friends with Smartphone Savi. His job is to teach kids how to choose the best apps to download.

He also shows them how to use apps safely, so they can have fun while they learn (and their parents don't have to worry).

The first character introduced in the Cyber City book series is **Digital Dan**, the leader od the city,who takes the form of a cloud.

He represents the cloud storage system and teaches kids how the cloud works and how to stay safe online through engaging rhymes and pictures.

The second character introduced in the Cyber City book series is **Padlock Picasso,** a talking padlock who arrives in Cyber City with a suitcase and very special paintbrush.

She immediately excites the children by showing them her scroll of password rules that she's written with her paintbrush.

The third character is **Smartphone Savannah**, who goes by **Savi,** and takes the shape of a small, exciting smartphone.

She comes to Cyber City to teach kids how to use their smartphones responsibly. This is an essential skill for young learners, who may find themselves using smartphones at home, in school, or even in unsupervised situations as they get older.

Just like all the other wonderful teachers, Appi Alex wanted to show off his features.

He came to teach the
kids all about apps,
and play some games
for fun, perhaps.

Appi Alex showed up
very much alone.
He looked just like an
app on a mobile phone.

But Alex was very excited
and felt great.
To see his best friend
Savannah, he couldn’t wait.

Appi Alex's eager students somehow always knew, that apps were very interesting, and just not a few.

He told them these could
be their learning buddy,
if they used them not just
for play, but also to study.

LEARNING
1
2
3
4
A B D
C G
E F

He sighed and said it was
really a very big shame,
to use their apps only to
have fun or play a game.

Since they had to study in order to achieve their dream, so they should choose apps with lessons from STEAM.

STEAM stands for the subjects Science, Technology, Engineering, Art and Math.

All of which need no apology.

S
T
E
A
M
Science
Technology
Engineering
Art
Math

Choose one early to follow all the way through school.
Apps are there to help you learn, if you'll stick to the rule.

With any program or app, be careful to what you agree.

And don't be in a hurry to download everything you see.

DOWNLOAD

Choose only those apps which you really do need.

And use them to learn or find something to read.

SUN
ART
PRINCESS

Apps, as you know, can take
up quite a lot of space,
so the ones you don't use,
you should always erase.

Before something goes wrong
when you set out to buy,
always talk to your parents
before you even begin to try.

APP
APP
APP
APP
APP

Never buy or download without asking your parents first.

Don't lie. When they find out after, it will be the worst.

BILL
App
App
pp
App

You must listen to your parents, whatever they have to say.

Believe me, you will come to thank them for this one day.

APP STORE
SALE!
APP
APP
APP

Appi Alex was happy with
all that he shared,
because kids who listened
were well prepared.

Words to Know:

apps	(applications) programs designed to do a particular job
art	creative activity, such as painting, music, poems, and dance.
bully	a person who seeks to harm for frighten someone
cyber	a digital world of the online world of computer networks
Cyber City	a place where Smartphone Appi Alex and friends live and work
cybersecurity	things done online to make people and places safe
computer	a device that performs a task automatically
data	fact or information stored on a computer
download	to get data from another computer, usually using the internet
erase	rub out or remove (delete)
engineering	to design and build (a machine or structure).
features	something important or interesting
information	what you know about something or someone
internet	something that connects computers throughout the world
math	the science of numbers, amounts and shapes
permission	give agreement for something to happen
science	knowledge and understanding of events in nature and space
STEAM	Science - Technology - Engineering - Art - Math
Technology	the study and use of tools and machines made by people

Made in United States
Troutdale, OR
11/03/2026

56845109R00024